AF350632

The Cambrian Period: The History and Legacy of the Start of Complex Life on Earth

By Charles River Editors

A picture of Archeocyathids fossils found in California

About Charles River Editors

Charles River Editors provides superior editing and original writing services across the digital publishing industry, with the expertise to create digital content for publishers across a vast range of subject matter. In addition to providing original digital content for third party publishers, we also republish civilization's greatest literary works, bringing them to new generations of readers via ebooks.

Sign up here to receive updates about free books as we publish them, and visit Our Kindle Author Page to browse today's free promotions and our most recently published Kindle titles.

Introduction

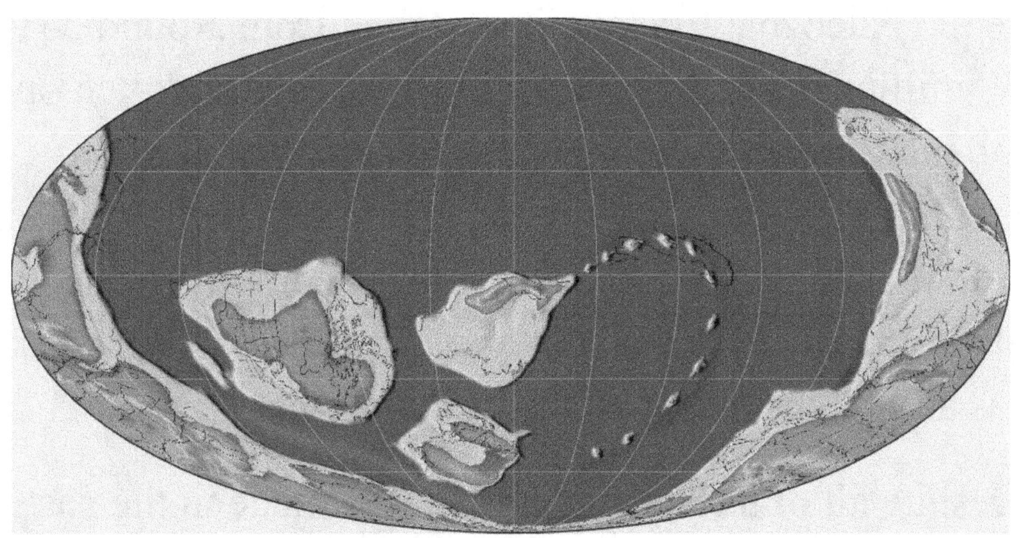

A map of the continents during the Cambrian Period

The early history of Earth covers such vast stretches of time that years, centuries, and even millennia become virtually meaningless. Instead, paleontologists and scientists who study geochronology divide time into periods and eras.

The current view of science is that Earth is around 4.6 billion years old, and the first 4 billion years of its development are known as the Precambrian period. For the first billion years or so, there was no life in Earth. Then the first single-celled life-forms, early bacteria and algae, began to emerge. It's unclear where they came from or even if they originated on this planet at all, but this gradual development continued until around four billion years ago when suddenly (in geological terms) more complex forms of life began to emerge.

Scientists call this time of an explosion of new forms of life the Paleozoic Era, and it stretched from around 541-250 million years ago (Mya). In the oceans and then on land, new creatures and plants began to appear in bewildering variety, and by the end of this period, life on Earth had diversified into a myriad of complex forms that filled virtually every habitat and niche available in the seas and on the planet's only continent, Pangea.

Despite all of the scientific advances made in the past few centuries, including an enhanced understanding of Earth's geological past, very little is known about the planet's early history. It is generally accepted that the planet formed somewhere in the region of 4.5 billion years ago, and at some point, the first life appeared in the form of tiny, single-celled creatures, but scientists are unsure of what this life looked like. One of the problems for those seeking to trace the history of life on Earth is that modern scholars are almost entirely dependent on fossil records, but the earliest types of life left few fossils. The best fossils are formed from the bones and hard body parts of dead creatures, but the earliest types of life were so small that they had no bones or cartilage and thus left no fossils. Thus, even though the Precambrian Period (4,600–541 millions of years ago (Mya)) covers over 80% of the entire history of the planet, scientists have very little idea of what forms of life existed then.

The Precambrian Period is divided into three eons: the Hadean (4,600–4,000 Mya), the Archaean (4,000–2,500 Mya) and the Proterozoic (2,500–541 Mya). One layer of carbon in rocks that date to the Archaean Eon, found on islands west of Greenland, seems to have fossilized tracks that may have been left by some form of organic life, but scientists simply are not certain. The oldest confirmed remains of life date from the same eon and were found in rocks in Western Australia, which were found to contain fossilized bacteria estimated to be 3.46 billion years old. It seems likely that there were thousands, perhaps even millions of forms of life in these early periods, and most seem to have been microscopic in size and very simple in developmental terms.

During the Proterozoic Eon, the first forms of multicellular life began to appear. Evidence of one of the earliest possible examples of multicellular life was found in rocks dating from around 1,500 Mya. Another example of multicellular life was deposited in the Lantian Formation dated at about 580 Mya.

Soft-body life forms, like jellyfish, appear to have been quite common throughout the world from 635 to 542 Mya. These are called the Vendian or Ediacaran biota. Creatures with hard shells also began to appear towards the the end of the Ediacaran Period and the Proterozoic Eon.

Then, as Earth entered the Cambrian Period, there was a relatively sudden increase in life form diversity throughout the oceans. Completely new forms of life, more complex and more diverse than anything that had been seen before, began to spread. This acceleration in the evolution of new forms of life was so dramatic that this has come to be known as the "Cambrian explosion."

Although new species in the Cambrian explosion developed almost entirely in the oceans, the land was not entirely devoid of life. Though there were no plants or animals, mats of cyanobacteria and other types of microbes covered large terrestrial areas. Scientists have discovered the tracks of a creature that were left in mud that existed 551 Mya, and those tracks were left by leg-like appendages. Was this a fish-like creature that temporarily invaded the land, or was it something completely different than anything that exists today? There is no general consensus, but the Cambrian Period left a rich fossil record that provides a clear idea of the development of life during this time. At the same time, new discoveries are continually being made, and the more scientists discover about this mysterious period, the more their understanding of ancient Earth changes.

The Cambrian Period: The History and Legacy of the Start of Complex Life on Earth looks at the development of the era, the extinction event that preceded it, and how life began to evolve during it. Along with pictures depicting important people, places, and events, you will learn about the Cambrian Period like never before.

Background

The modern understanding of the creation and development of planet Earth really began to emerge in the early 19th century when geologists first began to analyze rock strata and to recognize that different layers contained different kinds of fossil remains. It was quickly realized that these provided a snapshot of life at various periods in the ancient past and it didn't take long before scientists began to use agreed terms to describe the vast stretches of time that comprise the history of our planet.

One of the first was a self-taught English geologist named John Phillips and in the 1830s he published a book that would change our understanding of the history of the planet and finally standardize the terminology used to describe these ancient periods. He did this by ordering rock strata according to the different types of fossils found within them and used this to define different periods of the development of life on Earth.

Phillips

These classifications describe the history of life after the Precambrian period in terms of three eras, with each being further subdivided into several periods. These eras are:

The Paleozoic Era (meaning the era of "ancient life") was the oldest and a period of dramatic upheaval and change. It covers a period from five hundred and fifty million to two hundred and fifty million years ago. This era began with the emergence of the first multi-celled life and by its end, the first large reptiles, creatures such as Dimetrodon and Edaphosaurus, had appeared on the single continent, Pangea. This era is further subdivided into six geologic periods; Cambrian, Ordovician, Silurian, Devonian, Carboniferous and Permian.

The Paleozoic Era was generally a period of unbroken evolution of increasingly complex forms of life. However, it also included at least three of what have become known as "mass extinction events." For reasons that are not fully understood but are most likely associated with some form of climate change, there were three occasions during this era when many species completely died and were replaced by other forms of life.

The first of these events occurred between the Ordovician and Silurian periods. The Earth grew colder, glaciers appeared and sea levels and sea temperatures dropped dramatically. Around 25% of all species on Earth were killed. In the oceans the effect was even more dramatic and around 60% of all marine species vanished forever.

At the end of the Devonian period there was a second and even more catastrophic extinction event that led to the death of around 70% of all species on Earth. No-one is quite certain what caused this event though current theories include excessive sedimentation of the oceans, a period of rapid global warming (or cooling), the impact of a comet or large meteorite or even habitat changes caused by massive nutrient runoff from the land.

Then, at the end of the Permian period there was an extinction event so cataclysmic that it became known to early paleontologists as "the great dying." In a period that may have been a short as twenty thousand years, 95% of all species of animal were wiped out and virtually all trees and many plants disappeared. Just like other extinction events, no-one is entirely sure what caused the great dying, but recent discoveries may give a strong clue.

Rocks in present-day Australia and Antarctica have been discovered that contain tiny quartz crystals marked with microscopic fractures. Quartz is incredibly strong and it would take enormous force to do this, many times the power of a nuclear explosion. It is thought that perhaps a huge asteroid more than three miles (4.8 kilometers) in diameter may have struck the Earth causing a sudden and cataclysmic change in climate.

Clouds of particles and gases would have blocked out the sun for months or perhaps even years. Global temperatures would have initially dropped and corrosive rain and snow would have begun to fall. When this eventually stopped, the atmosphere would have been filled with greenhouse gases and a period of global warming would have followed that may have lasted for millions of years. Earth became a blank slate, ready for entirely new forms of life to appear.

The next era was the Mesozoic Era (meaning the era of "middle life"), spanning a period from around two hundred and fifty to sixty-six million years ago. Like the previous era, this was a period of continual change. It was also the period when large reptiles first began to appear and then became the dominant form of life on Earth. This era is subdivided into three geologic periods; Triassic, Jurassic and Cretaceous.

The Triassic period saw new forms of life appear on Earth to replace those killed during the Permian extinction. The first true dinosaurs began to appear, though they were much smaller than those that followed, and these spread out across the single continent, Pangea. However, towards the end of the Triassic period there was yet another global mass-extinction that saw around 35% of all species being killed. Paleontologists are not certain what caused this event but, like earlier mass-extinctions, it is thought to have been caused by some form of climatic change.

The most prevalent theory is that there was a period of large-scale release of greenhouse gases into the atmosphere, perhaps caused by widespread volcanic activity. This is thought to have caused widespread global warming and the acidification of the oceans. The rise in temperature caused the extinction of many species, but one type of creature seems to have been virtually unaffected: reptiles. Perhaps the cold-blooded reptiles were better able to survive rising temperatures? No-one is certain, but we do now that at the end of the Triassic period, the extinction of many other species left reptiles able to proliferate and expand to fill virtually every evolutionary niche.

The following Jurassic period saw the beginning of complete reptile dominance of the planet. Huge herds of sauropods roamed Pangea's fern-covered interior, hunted by predators such as Allosaurus. In the Oceans, mighty marine reptiles, the Mosasaurs, hunted fish and other sea creatures. Flying reptiles, pterosaurs, dominated the skies. By the end of the Jurassic period, dinosaurs were the most successful and diverse form of life on earth. In the period that followed, the Cretaceous, they would become even more abundant, diverse and specialized.

There are no precise dates associated with the Jurassic period. The spans of time involved are so vast that all we have are vague approximations. For example, the end of the Triassic period and the global extinction event that followed may have covered a period of twenty thousand years. Or two hundred thousand. Or two million years. Each has been suggested and each has been accepted (and rejected) by some paleontologists. The date given for the end of the Jurassic period varies by as much as five million years according to different theories. It is important to remember that we are not dealing here with the kind of precise dates that characterize human history. The early history of planet Earth proceeded in a relatively leisurely way, with events unfolding over millions of years. The periods and dates we are dealing with here are neither fixed nor generally agreed with absolute certainty.

The most recent chapter in the history of the planet, and the one that modern societies live in today, is the Cenozoic Era, the era of "new life."

The following is the list of geological periods and the range of ages in millions of years ago (Mya):

- Precambrian - 4,600–541 Mya

- Cambrian - 541–485.4 Mya

- Ordovician - 485.4–443.8 Mya - Ending with first major extinction.

- Silurian - 443.8–419.2 Mya

- Devonian - 419.2–358.9 Mya - Ending with second major extinction.

- Carboniferous - 358.9–298.9 Mya

- Permian - - 298.9–251.902 Mya - Ending with third major extinction.

- Triassic - 251.9–201.3 Mya - Ending with fourth

major extinction.

- Jurassic - 201.3–145 Mya

- Cretaceous - 145–66 Mya - Ending with fifth major extinction.

- Paleogene - 66–23.03 Mya

- Neogene - 23.03–2.58 Mya

- Quaternary - 2.58 Mya to Present

The Cambrian Period is therefore both the first period of the Phanerozoic Eon and the Paleozoic Era. This is a particularly important period because it marks the beginning and sudden expansion of diverse, multicellular life on Earth.

In the Precambrian Era, there were simple life forms that evolved, but there are huge gaps in knowledge due to lack of fossils, so scientists must think like detectives, using the available evidence to deduce what may have happened. For example, from 2,000 to 1,000 Mya, tiny and microscopic life forms showed an increase in spiny outer coverings, suggesting that they had greater need for defenses to ward off predators. However, there is no way to determine what those predators may have been like because they left no fossil evidence.

From 2,700–1,250 Mya, fossil remains indicate there

were large numbers of stubby pillars constructed in shallow water by colonies of microscopic organisms. These pillars are known as stromatolites, but suddenly the diversity and abundance of stromatolites sharply declined. Some scientists attribute this decline to other types of life that disrupted the stromatolites by burrowing and grazing within those pillared communities, but if so, there is no way to know what those creatures may have looked like.

The existence of some life forms can be deduced not from their remains but from other evidence of their movements, and these are known as "trace fossils." For example, scientists have discovered strange trails in fossils of 565 million-year-old microbial mats. From this they have deduced that worm-like (vermiform) creatures moved underneath the mats and left trails, though nothing remains of these creatures.

The Ediacaran Period was the last period of the Proterozoic Eon and the end of the Precambrian. At the beginning of this period, a large percentage of existing acritarch life forms suddenly went extinct. These included a diverse array of organisms, ranging from small metazoans (multicellular eukaryotic organisms) to elements of many kinds of chlorophyta (green algae). These organisms had persisted for hundreds of millions of years with very little change, but during the Ediacaran Period, most disappeared and were replaced with a larger set of very strange creatures that had a degree of diversification not seen before.

These new creatures that emerged and thrived for the last 40 million years of the Ediacaran Period are collectively known as the "Ediacaran biota." These creatures evolved through three discrete stages, growing in complexity and size with each one, and a large portion of these life forms were completely different from anything seen on Earth before or since. Some looked like floating quilted mattresses, and others resembled mud-filled bags or simple disks.

Some of these strange forms may well have been behind the sudden surge in diversity known as the Cambrian explosion, either evolving into later forms or acting as catalysts for the evolution of others. Some of these forms may have been early versions of arthropods, echinoderms or even mollusks. Taxonomic classifications prove difficult because flora and fauna have been grouped by similarities in outward appearance and internal structure, but the Ediacaran biota have very few similarities with earlier or later forms of life, so attempts at classification remain tentative at best.

In fact, these very strange organisms are so mysterious that even their relationship with the diversity of the Cambrian Period is not fully understood or agreed on. The only consensus is that during the Ediacaran Period, some new life forms were beginning to develop in the oceans, while on land, there was little or no multicellular life before the beginning of the Cambrian Period. Vast areas were covered by dense mats of single-celled photosynthesizing organisms, but there seem to have been no complex organisms.

Nonetheless, as the Ediacaran Period drew to a close, Earth was poised to become home to all kinds of new organisms that would transform the planet.

The Geology and Climate of the Precambrian Period
There are relatively few geological clues as to what the early Earth looked like. Geology has left poor records of the history of tectonic plates during the Precambrian, though most scientists agree that this was a time of great movement on the crust of the Earth. The most recent theories of the Precambrian Period suggest that Earth had amongst its large oceans a number of small, nascent continents before 4,280 Mya, which was around 500 million years after the planet was first formed.

Evidence is scant from those far earlier periods, but from the little evidence collected, the first supercontinent may have been formed around 3,636 Mya. This first supercontinent is now generally known as Vaalbara, and the best estimates place the breakup of Vaalbara at 2,845–2,803 Mya.

Approximately 2,720 Mya, another supercontinent, Kenorland, was formed from that early breakup, and its own breakup occurred roughly 2,450–2,100 Mya. This split gave Earth what geologists call "proto-continent cratons," or blocks of continental material that were somewhat thicker than the oceanic crust surrounding them. These early "island" continents are today referred to as Laurentia, Baltica, Yilgarn and Kalahari, and when these cratons merged, they formed the supercontinent called Nuna 2,060–1,820 Mya. Nuna broke apart roughly 1,500–1,350 Mya.

Approximately 1,130–1,071 Mya, Rodinia formed out of the pieces cast off from Nuna. Some believe this supercontinent eventually broke apart roughly 600 Mya, forming eight individual continents, but other scientists have proposed that a "daughter" supercontinent formed almost immediately (in geological terms) from the breakup of Rodinia about 650 Mya and itself broke apart about 560 Mya, roughly 20 million years before the beginning of the Cambrian Period. This "daughter" supercontinent, if it existed, would mean that the breakup of Rodinia would have to have been several million years earlier than the current rough estimates. Like most everything in geology, beginnings and endings are usually not clearly defined or universally agreed upon.

The Precambrian Period is also thought to have featured several ice ages during which glaciers spread from the poles and the average temperature on Earth dropped dramatically. One ice age lasted for 300 million years, from 2,400–2,100 Mya, but the most intensely studied and best-known of the Precambrian Ice Ages is the Sturtian-Varangian glaciation, which lasted from 850–635 Mya. Some scientists believe that during this period of intense cooling, ice may have reached as far as the equator, creating a condition that has become kn0wn as "Snowball Earth."

The composition of the Precambrian atmosphere is also the subject of a great deal of debate and a degree of educated guesswork. The majority of modern geologists think that Earth's atmosphere started out as mainly nitrogen and carbon dioxide, along with a few inert gases. There is some evidence suggesting that the atmosphere may have had abundant quantities of free oxygen by the early Archean Eon, about 3,500 Mya, but the consensus view is that notable quantities of free oxygen did not exist in the Earth's atmosphere until life forms using photosynthesis began to develop and to convert huge quantities of carbon dioxide into molecular oxygen as part of their internal metabolic processes. As these new forms of life became established, the atmosphere evolved from being relatively inert to becoming chemically reactive.

This new, oxidizing atmosphere quickly gave rise to an ecological disaster sometimes referred to as the "Oxygen Catastrophe." This process would have been slow at first as free oxygen combined with other elements on the Earth's surface or on the sea floor. Iron, for instance, would have begun to rust for the first time as it combined with oxygen, and as long as there were plenty of raw oxidizable elements, most molecular oxygen would have been quickly taken out of the atmosphere. However, once the surface became saturated with oxidized minerals, oxygen began to build up in the atmosphere toward modern levels. Scientists have unearthed evidence of this oxygen buildup by finding giant formations of banded iron with layers of iron oxides.

The presence of oxygen in a planet's atmosphere is not a natural occurrence - in fact, astronomers have proposed methods for detecting free oxygen in the atmospheres of extrasolar planets as a way to determine if a world has photosynthetic plant life. Earth started with virtually no free oxygen in its atmosphere, so what many people and animals breathe today came from tens of millions of years of photosynthesis, with plants converting CO_2 into O_2 and animals converting oxygen, in air and dissolved in water, back into CO_2.

The earliest photosynthesis was done by cyanobacteria, but the oxygen they produced did not build up in the atmosphere to any great extent because free oxygen reacts quite readily with metals and the free oxygen in the atmosphere was initially absorbed by metals and other reactive compounds. However, once these materials had absorbed as much oxygen as they could, oxygen started to build up in the atmosphere, steadily replacing carbon dioxide until a new equilibrium was reached. Plants, animals and fungi need oxygen, and eukaryote diversity seems to be directly linked to the increase of atmospheric oxygen in the Precambrian world. Creatures also became larger as oxygen became more abundant, which makes sense since larger creatures require more oxygen. The biota of the Ediacaran Period grew to be several meters in length during the Precambrian Period.

In addition to the increasing levels of oxygen in the atmosphere that characterized the Precambrian Period, it is thought that the Earth also developed the Ozone Layer during this period, another essential prerequisite for diverse life. Without ozone in the atmosphere, life easily dies because sunlight includes potentially deadly ultraviolet radiation. Life in the deep oceans remains relatively protected, but life on land and near the surface of the oceans can be killed by high-energy UV light.

With higher concentrations of oxygen in the atmosphere, UV rays are absorbed by oxygen, energizing the molecule to combine with a third oxygen atom and produce ozone

(O_3). Lightning can also create ozone. Ozone is even better at absorbing ultraviolet light, effectively protecting the surface of the planet.

Most scientists believe that the ozone layer was already established before the Cambrian Period. This protection from UV light may have aided in the development of not only life on land, but also of more complex life in the shallow parts of the oceans.

The final significant factors during the Precambrian Period were the periods of heavy glaciation. The concept of "Snowball Earth" remains highly controversial and widely debated, but there seems to be a strong and growing consensus that a period of extended glaciation affecting much of the planet occurred during the early Ediacaran Period. Scientists theorize that such a period of heavy glaciation could have generated a mass extinction, essentially wiping the slate clean and leaving a very low level of diversity amongst the species that survived. Studies of subsequent mass extinction events such as those during the Ordovician-Silurian, Devonian-Carboniferous, Permian-Triassic, and Triassic-Jurassic eras indicate that life on Earth seems to rebound quickly from such events and new diversity often follows as creatures develop to fill new biological niches. The Ediacaran biota may have been examples of that rebound after a mass extinction.

On the other hand, some scientists think that the "Snowball Earth" event occurred far too long before the beginning of the Cambrian Period to have had any direct effect on the sudden diversity of life forms during that time. In fact, some point out that the period of extended cold may have hampered the development of larger life forms.

Evidence from recent research suggests that the oceans suddenly gained an influx of calcium towards the end of the Precambrian Period. Heavy volcanism from divergent tectonic plate boundaries may have created this surge in calcium content in the ocean. As an alternative to the Snowball Earth theory, some scientists have noted that when organisms have a new resource, sometimes they put that resource to use. This increase in calcium may have subsequently been the source of internal skeletons and external shells, but regardless of whether there truly was a "Snowball Earth" or not, there were certainly massive changes taking place on the planet that set the stage for the Cambrian explosion.

The Geology and Climate of the Cambrian Period

Throughout its history, Earth has never stopped changing, and for billions of years it has alternated between supercontinents and separate "island" continents. Tectonic movement caused the continents of the world to crash into one another, melding into one gigantic landmass, a supercontinent. Scientists still don't understand why such a massive single continent would then break up, but eventually something would send fracture blocks of continental material moving steadily away from each other. At the same time, since the materials were constrained to moving on a spherical surface, eventually they collide once again, forming a new supercontinent.

By the beginning of the Cambrian Period, it is believed that the supercontinent Rodinia, which had formed from the pieces cast off from the previous supercontinent (Nuna), had broken up into at least eight separate continents. The breakup of Rodinia occurred around 560 Mya, approximately 20 million years before the beginning of the Cambrian Period, but as mentioned before, some scientists believe that at least some of the eight individual continents may have come together before the start of the Cambrian Period to form a single "daughter" continent, Pannotia. If Pannotia existed, then it too had started to break up by the beginning of the Cambrian.

By the Cambrian Period, Siberia, Baltica, and Laurentia (North America) had already become distinctly separate landmasses. The greater percentage of continental material was huddled south of the equator during this time, although the majority appeared to be drifting northward. Some geological evidence suggests that one of the continents, Gondwana, experienced a geologically brief period of rapid rotational movement with respect to the other continents during the Cambrian Period.

It would not be until 150 million years after the end of the Cambrian Period that the eight continents would come together again to form the last supercontinent, Pangaea, around 335 Mya. About 175 Mya, during the Jurassic Period, Pangaea itself started to break apart, and that breakup is still ongoing, with the African tectonic plate tearing off a corner, now called the Arabia plate, about 35 Mya. The Somalia sub-plate remains an ongoing break even today, splitting the bulk of Africa from East Africa along the Great Rift Valley.

To understand the climate on Earth during the Cambrian Period, it is necessary to consider it within the wider context of climate changes over a geological timescale. Since the planet was first formed, the sun has been getting brighter by about 10% every billion years. This is due to the changing chemical nature within the body of the sun as it converts hydrogen into helium. Thus, from the formation of Earth to the start of the Cambrian Period, the sun increased in brightness by roughly 46%.

Furthermore, in the Precambrian Period, the atmosphere of the Earth was very different. Some studies suggest that in the distant past, CO_2 accounted for as much as 96.5% of the atmosphere, similar to conditions on Venus today. This would mean that the nitrogen content was closer to 3.5% of the atmosphere, and the total atmosphere would have been far thicker, yielding a surface pressure closer to 22.6 bar, compared to today's 1.013 bar. That would have made Earth far warmer then, despite the sun providing less heat.

Over hundreds of millions of years, the Earth's atmosphere reduced in thickness, resulting in a general cooling effect, and during that same stretch of time, the increase in the sun's energy output tended to increase the global temperature. These two factors, the thinning of the atmosphere and the increase in the sun's energy, virtually cancelled each other out, leading to a relatively steady overall average temperature. However, while the increase in solar output was steady and predictable, the changes in Earth's atmosphere caused by natural factors varied considerably over time. This led to times when the average temperature on the planet rose and times when it fell dramatically, leading in some cases to ice ages.

Studies have demonstrated that many of these ice ages follow a cycle of around 147 million years, coinciding with the Solar System passing through the heart of one of the Milky Way's galactic arms, where the cosmic ray flux is more dense. Changes in cosmic ray flux affect cloud nucleation rates and thus the albedo of Earth, making the planet cooler when rates are higher. A relatively nearby supernova could also send Earth into a new ice age outside of the galactic arm passage cycle.

During the Cambrian Period, Earth received about 4.7% less sunlight than today, and the planet cycled through a series of intense and extended periods of extreme cold, much as it had during the Precambrian Period. In the appropriately named Cryogenian Period, the planet experienced two extended ice ages. Precise dates are still a matter of debate and research, but it seems that the Sturtian Ice Age lasted from around 715–660 Mya and the Marinoan Ice Age from 650–635 Mya. Those two ice ages occupied virtually the entire Cryogenian Period.

During the Ediacaran Period, the last period of the Precambrian, Earth gradually recovered from these periods of extreme and prolonged cold, to the extent that by the start of the Cambrian Period, the Earth was on average around 45°F warmer than today, and that was considerably cooler than most of the periods that followed. Some of that coolness was likely caused by the continent of Gondwana extending over the south geographic pole, blocking warmer ocean currents from reaching it and allowing ice to accumulate.

Some scientists have found evidence of a short ice age they've called the Baykonurian, with their best educated guesses placing it about 547 Ma, only about 6 million years before the start of the Cambrian. Some suggest that this short ice age helped to contribute to the stresses which gave birth to the Cambrian explosion. The climate continued to warm throughout the Cambrian Period, gradually melting the Gondwanan ice cap and helping to raise sea levels significantly. Rising sea levels flooded great stretches of the continental interiors with relatively shallow, warm inland seas. These proved to be ideal for the development of new forms of life.

The Cambrian Explosion
The sudden, dramatic increase in diversity of many species of fauna during the Cambrian Period is typically referred to as the Cambrian explosion and sometimes the Cambrian radiation because a few species radiated outward to become many species. This is when most phyla of the animal kingdom first appeared in the fossil record over the course of about 13-25 million years, which spans about 25-50% of the entire Cambrian Period.

Before the Cambrian Period, most living creatures were microscopic, unicellular beings that were sometimes arranged in large colonial structures, somewhat like modern corals, but during the Cambrian Period, multicellular creatures developed and diverged to become markedly different species, each specialized to thrive in a particular environmental niche. Few of the creatures that first emerged during the Cambrian Period survive today, but this still represented the beginning of diversity in living creatures that continues to the present day.

The first understanding of the development of life during the Cambrian Period began with the discovery of trilobite fossils in the late 17th century, described in 1698 by Oxford Museum curator Edward Lhuyd. As a result of this discovery, about a century later, William Buckland (1784–1856) suspected that something remarkable had occurred at the stratum which now defines the early Cambrian.

A bust of Lhuyd

Buckland

In the 19th century, geologists like Roderick Murchison and Adam Sedgwick made use of the earliest fossils for the relative dating of rock layers, and by 1859, a great debate raged concerning the origin of life at what was then called the earliest stratum of the Silurian. Murchison and other leading geologists felt that they had found life's beginning, while Charles Lyell and many others disagreed with their assessment. This period was named Cambrian after the Roman name for Wales, which was where some of the earliest fossils from this period were discovered by British paleontologists.

Charles Darwin, then involved in an attempt to understand the origins of life on Earth, wrestled with this enigma and considered the sudden appearance of trilobites in the Cambrian Period to be "undoubtedly of the gravest nature." Darwin believed that evolution, the gradual development of species to take advantage of environmental niches, accounted for all life, so the seeming appearance of an entirely new species with no obvious antecedents was a major problem. Darwin suspected (correctly as it turned out) that the fossil record was less than perfect and far from complete. If Darwin was right, the seas of the Precambrian Period must have been rich with life that gradually evolved into more complex creatures such as trilobites, but he struggled to find fossils that could prove the existence of these earlier life forms. In one of his later editions, Darwin emphasized the importance of this problem and conceded, "To the question why we do not find rich fossiliferous deposits belonging to these assumed earliest periods prior to the Cambrian system, I can give no satisfactory answer."

Darwin

One rather creative American paleontologist, Charles Walcott, suggested that the period before the Cambrian did have many forms of life, but that they did not form fossils for some unknown reason. He gave this supposed period the name "Lipalian" and theorized that the animals which show up in the Cambrian fossil record had evolved during his hypothetical earlier period.

Walcott

In fact, part of the problem was that early paleontologists were unaware that much of the life that existed before the Cambrian Period consisted of microorganisms. These did leave fossil remains, but they were not nearly as obvious as the larger trilobites and other creatures that characterized the Cambrian Period. Currently, the earliest fossil evidence is that found in Warrawoona, Australia, dating to 3,850 Mya. Scientists claim that colonies of microorganisms which formed stubby pillars called stromatolites were found in fossil form there.

Later fossil evidence found in both China and Montana, dating to 1,400 Mya, contained significantly more intricate life forms from which all fungi, plants, and animals subsequently evolved. Though Darwin had hinted at the existence of a Cambrian transition period when he discussed the imperfect nature of the fossil record, it wasn't until 1948 that American paleontologist Preston Cloud was able to describe a phase of "eruptive" evolution near the beginning of the Cambrian Period, during which life rapidly developed from single-cell to multicellular forms. At the time, his thesis was really no more than an educated guess, and it wasn't until the 1970s that scientists were able to find evidence to back it up.

British paleontologist Harry Blackmore Whittington led the way to a new understanding not by discovering new fossil deposits but by re-examining existing finds, especially those of the Burgess Shale, consisting of fossils found in an area of the Canadian Rocky Mountains known as the Burgess Pass. These deposits were part of Laurentia, one of the continents of the Cambrian Period, and they are recognized as some of the most diverse fossil finds in the world. In order to protect the location of these fossils from eager paleontologists and amateurs, in 1981 UNESCO designated the Burgess Shale a world heritage site.

Whittington realized that these deposits included fossils of multicellular life forms, including the most common for that stratum, Marrella, an arthropod that was unlike any other arthropod class known to scientists at that time. Many of the creatures discovered by Whittington and others were very odd indeed and clearly have not survived to the present. These included the Wiwaxia, a slug-like, spiny creature, and Opabinia, which possessed five eyes. In 1985, Whittington published *The Burgess Shale*, a book which described his finds and brought the notion of the Cambrian explosion a measure of scientific acceptance.

In 1989, American paleontologist Stephen Jay Gould published *Wonderful Life*, another book based on finds in the Burgess Shale that became a bestseller and further popularized the notion of the Cambrian explosion. However, Gould also raised important issues based on these fossil deposits that questioned the very fundamentals of how evolution really worked. Darwin postulated that evolution must progress slowly, with steady and gradual development, but Gould argued for the concept of "punctuated equilibrium," the idea that evolution actually happened in relatively sudden spurts in response to environmental changes. Gould pointed out that the fossil record from places like the Burgess Shale contains very little to support the gradualism asserted by Darwin. Instead, the evidence in the ground shows that species persisted for millions of years with very little change, then suddenly transformed or speciated in a relatively brief window of time.

While Gould came later and was working with a more complete fossil record, it's important to keep in mind that the record remains incomplete, especially when it comes to early forms of life. The simple fact is that not every species left a fossil record, and among those that did, some left only one fossil or a few. It is a virtual certainty that millions or perhaps even billions of species of animals lived and died without leaving a trace that they had ever existed.

Even without being able to ascertain how many different species of life there were in the Precambrian Period, it is still generally accepted that the evolutionary pace did increase by an order of magnitude during the time described as the Cambrian explosion. Despite that consensus, however, there is plenty of debate over the specific cause of this sudden surge in diversity.

Most scientists cite two broad categories of causes for this unique evolutionary event:

1. Ecological and Environmental factors,
2. Developmental factors.

While the effects of increasing oxygen in the atmosphere, the appearance of the ozone layer, and a possible mass extinction event caused by "Snowball Earth" have already been discussed, there are other possible ecological factors that may have impacted the Cambrian explosion.

One is the impact that new species may have had on the environment in which they existed. During the later Devonian Period, for example, when life colonized the terrestrial world by crawling out of the oceans for new opportunities, the changes on land led to changes in the oceans. For example, the creation of new sediments on land led to new patterns of nutrients in the ocean, and that change had a huge impact on marine life across the world. There is no direct evidence of such a change in the Cambrian Period, but some scientists have speculated that the appearance of so many new species in such a short time may have dramatically changed the ecology of the planet.

It also seems likely that smaller changes in the ocean occurred where the actions of one new species made life suddenly viable for another entirely different species. For example, burrowing creatures aerate the sediment on the bottom of shallow seas, making available nutrients and other resources that were previously buried. Sponges in the benthic zone filter tiny particles from sea water and place them in the mud below, providing potential nutrients for new species. As the actions of a new species impacted the ecology of the area in which it lived, this in turn created new opportunities for diversity.

The Snowball Earth theory is also not the only possible trigger for a Precambrian mass extinction event. Scientists have uncovered evidence of the existence of near-simultaneous anoxic events around the world, and these have been linked to the sudden disappearance of shelly fossils (like Cloudina) and many species of Ediacaran biota shortly before the Cambrian. Studies of other mass extinction events that came later show that when ecological niches are made empty by extinctions, surviving clades scramble to fill those empty ecospaces by developing new diversity.

On the other hand, one developmental factor that has been cited as a possible driver for the Cambrian explosion is the evolution of eyes. Before the evolution of eyes, marine creatures would have been aware of something in their immediate environment by touch, chemical interaction similar to smell, or even by vibrations in the water. But when a predator could see its prey at a distance, the situation became very different, and the relatively defenseless prey needed to rapidly evolve a new set of defenses, such as armor (shells) or spines.

Researcher Andrew Parker has studied this dynamic as a possible cause for radical changes in the diversification of life during the Cambrian explosion. Some of the fossils found of creatures from the Cambrian Period are the earliest known to have had eyes, and Parker further pointed out that in environments where sight is impossible, such as the deep ocean or in caves, creatures tend to lose some of their diversity.

Other scientists disagree with Parker about eyesight being a major factor for the Cambrian explosion. Some suspect that vision developed far earlier, though there is no clear evidence to support that belief yet.

Although there is no general agreement about the importance of eyes or when organisms began to develop eyes, most scientists do agree that the early Cambrian Period was characterized by a prolonged and rapid "arms race" between predators and prey which drove specialization. The life or death nature of the predator-prey situation remains one of the most powerful parts of natural selection, and it has been accepted as a fundamental driver of evolutionary change as far back as Darwin's day. Scientists readily acknowledge that the impetus to adapt is far stronger with the hunted than with the hunter, which makes sense given that the hunter can fail on occasion, while the hunted must never lose.

The critical point in any discussion on this dynamic involves awareness, especially on the part of the hunted. How are the prey aware that they are being sought? Awareness, intelligence, witnessing attacks, and some biochemical drive to avoid attacks would seem to be involved. Without such awareness, intelligence, and motivation, even on the most rudimentary level, there is no reason to believe life would evolve in a useful or more complex direction. Random changes would not take place fast enough, and Gould's idea of "punctuated equilibrium" argues against a purely accidental path toward such utility.

Predation as a method for acquiring sustenance likely started long before the Cambrian. Acritarchs became increasingly spiny, a development that would only make sense if it was for protection against predators. Something drilled holes in Cloudina shells, among other evidence of prey being hunted, and while it is unlikely that predation alone drove the Cambrian explosion, the establishment of food chains may have had a profound effect on the direction of the period's evolutionary processes.

One other factor to consider involves the intensity of predation. There is evidence that it increased quite strongly throughout the Cambrian Period as predators evolved new strategies for gaining their next meal, including the ability to crush the shell of their prey.

One reason scientists know the Cambrian explosion was due to factors other than predation is the genus-level pattern across time of median predator ratio. Throughout both the Cambrian and Ordovician periods, there is no strong correlation between diversification and the predator ratio.

Co-evolution added another dynamic to the evolution of metazoans (animals). The traits in one species can lead to other traits being produced in other species, and in that way, for any given threat, there may be a number of possible responses. For example, prey might respond to a predator's new tactics by evolving skills for avoidance or fleeing, or alternatively by developing new skills of defense. With this bifurcation of talents, predators may be compelled to evolve to pursue both, and through speciation, one group may evolve the ability to break through defensive measures while another group develops the ability to more quickly pursue prey.

Lastly, some scientists think the Cambrian explosion was not a huge evolutionary event, but merely representative of a boundary being crossed, beyond which far more possibilities were suddenly available. With very little oxygen available, organisms were constrained to a limited set of tools, but once oxygen became abundant, life forms had many more possibilities for diversity, and the Cambrian explosion was nothing more than the "moment" in time when many of those possibilities were suddenly realized. Crossing a genetic threshold, life could start to rely on an entirely new set of diversification strategies. Anaerobic processes produce far less energy than a metabolism driven by oxygen use.

For example, oxygen breathers can create far more complex proteins, each of which open new pathways for even more complex evolution. Such proteins are like a new, larger toolkit which gives life the ability to evolve a far broader array of problem-solving strategies. With such a toolkit, organisms had many more possible ways to evolve, and sure enough, from a single-celled eukaryote, life on Earth eventually evolved into millions of species of animals, plants, and fungi of many shapes and sizes. Mushrooms, pineapples, and dolphins all came from the same ancestor, and the Cambrian explosion may well have been the open gate that made it all possible.

Cambrian Life Forms

Before discussing the creatures that appeared during the Cambrian explosion, it is important to understand some of the terminology involved.

A phylum is the highest level of classification for life forms on Earth. A phylum would be, for instance, a group of animals that share the same overall body plan. However, some animals are grouped under one phylum despite having wildly different outward appearances, because they are classified by internal similarities that enforce the notion that they are likely related.

Some might wonder why whales and giraffes are related, and it all has to do with the internal structure. Scientists suspect that the ancestors of both were terrestrial creatures and some of their offspring went back into the oceans, becoming whales and dolphins, while others remained on land, becoming giraffes.

In the animal kingdom, there are 34 known phyla. One of those is phylum Chordata, consisting of creatures with a spinal cord or dorsal nerve channel. Within this phylum, there are several classes, including fish, amphibians, reptiles, birds, mammals and the extinct non-avian dinosaurs.

Within the class Mammalia, scientists have well over a dozen taxonomic orders, including Primates (which contains humans) and Artiodactyla, which contains alpacas, antelopes, camels, cattle, deer, giraffes, goats, hippopotamuses, llamas, mouse deer, peccaries, pigs, and sheep.

By merely looking only at the outward appearance, one might look at tapeworms and earthworms and think that they are closely related because of their similar shape, but these two life forms occupy entirely different phyla.

With today's greater skill of genetic analysis, scientists are reworking their past ideas of existing phyla. The classification of life forms is still far from perfect. Many sources disagree on numbers and definitions, and many do a poor job of incorporating species that have gone extinct.

For most of the history of life form classification, scientists were concerned with immediate family members of a phylum, order, class, etc. To broaden the discussion, scientists have defined "stem groups" to refer to aunts and cousins of an immediate taxonomic clade. Stem groups are relations that originate with an ancestor prior to a family's earliest common ancestor.

Having this extra dimension in their conversations about Cambrian life forms makes it easier for scientists to discuss different possibilities.

Triploblastic refers to a common design found in many different types of life forms. This design consists of three key layers:

- Outermost - skin

- Innermost - gut (digestive tract)

- Middle - muscle and other internal organs other than the gut

Very few animals are not triploblastic - worms are, as are birds, fish, snakes, humans, and many others. Some of the creatures that are not triploblastic include jellyfish, sea anemones, and sponges. It is believed some of the earliest multicellular creatures may not have been triploblastic.

All animals that have left and right sides at any point in their development are considered to be bilaterian. Because of this left-right quality, such a creature necessarily has distinct bottom and top surfaces, and unique back and front ends.

All of the known bilaterians are triploblastic and vice versa. Some creatures, like sea urchins, sea cucumbers, and sea stars, may seem radially symmetrical, but they have bilateral symmetry in their earlier larval stage. Creatures like sea anemones, jellyfish, and sponges are radially symmetrical throughout their lives and thus don't fit into this category.

Coelomate is another term which refers to the overall design of a life form. This word refers to a creature which possesses a body cavity which will naturally contain the life form's internal organs.

Most of the Cambrian creatures were coelomates. Flatworms are an example that is not a coelomate because its internal organs are not surrounded by specialized tissues which form such a cavity.

Cladistics is another tool scientists use and involves their attempts to form a "family tree" for the different forms of life they have discovered. For example, if there are three groups of life forms, and the first two life form groups have far more similarities with each other than the third group, then they are considered to be closer on the family tree than either of them is to the third group.

Outward appearance can be misleading, but it is part of the set of clues scientists take into account for determining family trees. This can include the shape of a skeleton or exoskeleton, as well as the number of bones and their shapes. Scientists also consider things like DNA, if available, or the proteins within a creature's body.

Once a comparison is completed, scientists will have constructed a hierarchy made up of "clades," taxonomic groups which have members scientists feel ultimately have a mutual ancestor.

This cladistic technique isn't without problems, because life has a way of solving the same problem in very similar ways but in entirely different locales. Such solutions may have evolved numerous times, converging on the same or similar solutions. In plants, C4 species came up with similar solutions to carbon dioxide starvation when CO_2 levels dropped down to 800 parts-per-million (twice today's levels). The plants were not closely related, but developed C4 carbon fixation on their own, sometimes in ways very similar to other plant species.

By looking at the likeliest relationships, scientists can confine the possible range of dates for when a branch of the family tree first appeared. For instance, if fossils for two groups date to 20 million years ago, and scientists estimate that a third group was an ancestor of the other two groups, the third group is estimated to have evolved longer than 20 million years ago.

In the space of around 25 million years, a brief period in geological terms, the beginning of the Cambrian Period saw the emergence of most animal phyla with body plans that can be identified by fossil finds. Most of these new life forms were aquatic, inhabiting the warm, shallow seas that had formed as the ice melted and the average temperature rose.

Compared to earlier periods, scientists have a relatively good knowledge of the creatures of this period for two separate and distinct reasons. First, sedimentary deposits known as Lagerstätten were laid down around the world at this time. The word is German, meaning "storage place," and it is apt because these deposits contain some astonishingly well-preserved fossils that show not just bones and/or hard carapaces but also soft tissue. It is thought that anoxic conditions led to an absence of the bacteria that cause decomposition. The bodies of dead creatures were therefore preserved as they left impressions in soft mud that gradually became sedimentary rock. This contrasts sharply to fossils of dinosaurs, whose bones were often preserved but virtually all soft tissue has disappeared.

Lagerstätten deposits contain some of the most perfectly preserved fossil remains, and extensive deposits of this type dating from the Cambrian Period have been discovered in China, the United Kingdom, United States, Canada, Australia, and Scandinavia. The Burgess Shale discoveries were all found in Lagerstätten deposits.

The second reason there is a strong fossil record of this period is that creatures themselves were changing. Some life forms were developing hard outer shells and others became mineralized, meaning that they were more likely to leave fossil remains even outside Lagerstätten deposits.

Nonetheless, dating the beginning of the Cambrian Period has proved challenging for paleontologists. Most periods are defined by the appearance of a new and widespread form of life. Originally, the Cambrian Period was defined by the first appearance of trilobites in the fossil record, and in the early stages of paleontology, these were considered to be one of the first complex life forms. Later, however, scientists found fossils of other shelled species that had existed prior to the trilobites. Then the discovery of Ediacaran biota at a far earlier age led stratigraphy experts to demand a more precise definition for the Cambrian period base. After much searching and a false start, scientists settled on the Cambrian base being defined as the first appearance of *Treptichnus pedum ichnofossil*. This was some form of burrowing creature, but it is unknown precisely what it looked like because no fossil remains have been found. Instead, *Treptichnus pedum* has become known through the discovery of what are known as "trace fossils," fossilized remains not of the creature itself, but of the burrows it created. The lack of any fossil remains of the creature itself has led scientists to conclude that this creature lacked bones or any form of hard outer shell but the first appearance of *Treptichnus pedum* burrows in fossil form is now generally accepted as marking the boundary between the end of the Ediacaran Period and the beginning of the Cambrian Period.

Once this boundary was established, radiometric dates determined at locations from all around the world corresponding with this new Cambrian base allowed scientists to agree to a date for the beginning of the Cambrian Period. Early dating attempts hovered around 570 Mya, but the techniques were later found to be inaccurate and not suitable for this dating challenge. Finally, most scientists agreed on a date of 541 ± 0.3 Mya, and derived confirmation for this new base by the fact that earlier Ediacaran fossils ended just below that stratum.

However, some scientists have found problems with the site for the Cambrian base determination, claiming that the start of the Cambrian is more likely 544–542 Mya. Debate continues on this refinement.

Of all the types of fossil records, those of the full body of an organism including soft tissue have proved to be the most helpful in understanding a life form, and the Lagerstätten deposits created during the Cambrian Period have proved to be a rich source of such fossils. However, it is important to consider the abundant fossil record of the Cambrian Period in the context that the act of fossilization is very rare. Millions or even billions of examples of a species can live and die, and scientists may never find evidence they existed, and even amongst fossils that are created, many are destroyed by erosion or other kinds of natural damage, long before scientists ever have a chance to study them. As a result, the fossil record of life on Earth remains very incomplete, and the further back in time scientists dig, the more this tends to be the case, with rare exceptions.

Species which contain body parts that start out mineralized tend to have fossilized more often, like mollusk shells, though the soft portions of individual bodies tend to decay and disappear with time. Due to this, among the more than 30 animal phyla living today, less than 40% ever left a fossil record.

Scientists are lucky that the Cambrian contains an abnormally high quantity of Lagerstätten layers, which hold the shapes of soft tissue. Because of this, scientists have a better idea about the internal structure of animals from this critical period in the history of life on Earth. In most other types of sediment, scientists would only find things like claws, spines, shells and similarly hard portions of bodies.

Lagerstätten seem to preserve their captured organisms rather quickly, like in a mudslide, where the animal is buried alive, cannot escape and is rapidly suffocated. The conditions which make Lagerstätten possible are likely abnormal, so give a biased picture of the past, so they must be taken with other fossil records to gain a more complete, well-rounded picture of the past.

Scientists are still discovering and trying to understand fossil remains for the Cambrian Period. One of the most exciting Lagerstätten discoveries took place as recently as 2019, at the Danshui River, in Hubei, China. This has become identified as the Qingjiang biota. Within this Lagerstätten, the more than 200,000 fossil specimens from the Cambrian Period that have been collected include algae, sponges, arthropods, worms, sea anemones, jellyfish and other soft bodied animals. The soft tissue record from the Qingjiang biota was so incredibly detailed that researchers could see eyes, guts, mouths, gills, and muscles of creatures that lived 518 Mya. When scientists reported on their findings, they noted that virtually half of the identified species had never been seen before.

While it would be impossible to cover every new form of life that appeared during the Cambrian explosion, there were some common and notable new creatures that appeared at this time.

Some of the creaturess most associated with the Cambrian Period are Trilobites, an incredibly diverse and successful family that continued to live in the oceans long after the end of the Cambrian Period. Trilobites (the name means "three-lobed") were marine arthropods, invertebrates with an exoskeleton, a group that includes insects, arachnids, and crustaceans. However, though trilobites first appear in the fossil record from around 530 Mya, scientists cannot assume that they did not exist before that, because when the earliest trilobite fossils first appeared, they showed up all over the planet and in a range of diverse species. The inescapable conclusion from these facts is that trilobites had most likely already been around for quite some time and were likely already millions of years old by the time the oldest fossils were found.

For trilobites, the earliest fossil record comes from a time when they had developed mineralized exoskeletons. This may not be the earliest time they had exoskeletons, but it is possible that earlier forms had exoskeletons that were less susceptible to being fossilized.

Compared to anything that had come before, trilobites were complex and sophisticated creatures that developed to fill a number of environmental niches. Some appear to have crawled on the seabed, some scavenging, some filter feeding and some preying on smaller creatures. Some trilobites swam in the oceans, feeding on plankton. Some developed symbiotic lifestyles with other forms of life; the *Olenidae* family of trilobites are believed to have developed a relationship with sulfur-eating bacteria, which they ate.

Moussa Direct's picture of trilobite fossils

From the time that they first appeared, trilobites exploded across the oceans of the planet, exploiting a whole range of ecological and environmental niches. No one is entirely certain where trilobites came from. Similarities have been noted with fossils of Precambrian creatures including the Ediacaran fossils *Spriggina floundersi* and *Bomakellia kelleri*, both found in Australia, and *Parvancorina*, found in China and other locations. However, while these earlier creatures have similarities to the later trilobites, they also have significant differences. To the present, no fossil remains have been found that seem to be direct ancestors of the Cambrian trilobites.

10 different orders of trilobites have been recognized

and, within these, over 150 families, about 5,000 genera, over twenty thousand different species! Despite the range of species, all trilobites are basically similar with hard shells, multiple body segments and pairs of jointed legs and antennae. All trilobite fossils discovered have a similar body plan, with a head (cephalon), a segmented middle section (thorax) and a tail (pygidium). However, it is not these three parts that the trilobite name comes from. All trilobites have a long central axial lobe running from the cephalon to the pygidium. This central lobe is flanked by two pleural (side) lobes. These three lobes are common to every species of trilobite and are what gives this creature its name.

Trilobites are notable because they seem to have developed one of the first sophisticated visual systems of any creature. Most species have a pair of compound eyes (eyes comprising many individual lenses) on the outer edges of their heads. Compound eyes are particularly good at detecting motion and are still found in many current arthropods including spiders and insects. These give a significant advantage to a predator using eyesight to detect potential food sources. However, some trilobite species are eyeless. These are thought to have been bottom feeders which evolved to live on the deep ocean floor where light is poor and good eyesight does not confer a significant advantage.

Trilobites ranged in size from tiny examples little larger than a human fingernail to the massive *Isotelus rex*, which could reach over 700mm in length. The Cambrian explosion saw a vast number of different types of trilobites appear all over the world. By the end of the Cambrian Period, trilobites had reached the peak of their diversity and began a gradual decline in the total number of species after that time. However, trilobites continued well after the end of the Cambrian Period. They continued to flourish for three hundred million years and there were still a large number of trilobite species in existence up to the end of the Permian Period, around 251 Mya. At that point there was a mass extinction event that killed more than 90% of all species on Earth. No trilobites survived this extinction.

An *Isotelus* fossil

Although trilobites were the most common and best-known creatures to emerge during the Cambrian explosion, they were not alone in the shallow oceans five hundred million years ago. Another type of marine creature, Echinoderms, also began to appear. Echinoderms (the word is taken from Greek for "hedgehog skin") consist of a group of marine invertebrate deuterostomes that are the ancestors of today's starfish and sea cucumbers, amongst many others. Just like trilobites, echinoderms appeared relatively suddenly in the fossil record and have been found across the world.

Echinoderms are characterized by the presence of a plated calcite skeleton and tube feet used in feeding, movement and respiration. By the time that the first echinoderm fossils were created, there were already a great many diverse types and species. This means that these creatures must have existed before the beginning of the Cambrian Period, and the lack of earlier fossils has led paleontologists to speculate that the earliest variations may have lacked a calcite skeleton and were therefore less likely to leave fossil remains.

Unlike trilobites, which are broadly similar in appearance across the many species, the earliest echinoderms display an extraordinary range of sometimes bizarre body shapes. Many of these early species seem to have been relatively short-lived and scientists have speculated that these early examples were experimenting with different body shapes to find the best adapted to living in various environments. The most successful proliferated while others died out relatively quickly.

All the early forms of echinoderms lived on the seabed. The most common form found in the Cambrian Period were members of the now extinct class of *Eocrinoidea*, benthic suspension feeders with five ambulacra on their upper surface around the mouth and extending into a number of narrow feeding arms. Another form of echinoderm that first appeared during the Cambrian explosion were the *homalozoa*. Although they shared some features with *eocrinoidea*, *homalozoa* evolved into some very odd creatures indeed, displaying a bewildering variety of body shapes. Some fossil remains are so strangely shaped that paleontologists have been unable to decide which body opening was the mouth and which was the anus!

Unlike trilobites, echinoderms survived the mass extinction at the end of the Permian Period and survived to the present day. Today, there are around 7,000 different species of echinoderms living in the oceans of the world from the abyssal depths to tidal rock pools. These include starfish, sea cucumbers and sea urchins. All can trace their ancestry back to the creatures that first emerged during the Cambrian explosion.

Alongside echinoderms and trilobites, it seems that the Cambrian Period also produced the first true **crustaceans**, though the fossil records for these creatures are much less common. Fossils have been discovered in the Swedish Orsten formation from the late Cambrian Period that are unmistakably crustaceans, but these are all tiny, less than two millimeters in length. Scientists continue to debate whether these are miniature crustaceans or merely juvenile examples of late Cambrian crustaceans.

In the Canadian Mackenzie Mountains, at the Mount Cap horizons, more data that may confirm the presence of crustaceans in the Cambrian Period has been identified. Dating to the late Early Cambrian Period, the strata of rock, when dissolved with hydrofluoric acid, leaves behind microscopic remains including filtering appendages and mouthparts (mandibles) that unmistakably belong to some ancient form of crustacean. The diversity of these fragments is quite similar to that found in today's crustacean life forms. By analyzing all the pieces, scientists have been able to reconstruct a creature that seems to have fed using mandibles (one of the defining features of modern crustaceans) and may have been up thirty centimeters in length.

The Cambrian explosion also saw a dramatic increase in another type of creature, **cnidarians**. The phylum cnidaria is divided into two clades: *Anthozoa*, which includes corals and sea anemones, and *Medusozoa* which includes jellyfish. Little was known about the development of jellyfish in the Cambrian Period until the discovery of exceptionally preserved fossils in the Marjum Formation in Utah in the United States.

These fossils date from the Middle Cambrian Period, around 505 Mya and were created in very fine-grained marine sediments. This has provided fossils that not only show the soft body parts of these delicate creatures but even things like trailing tentacles. These jellyfish fossils look startlingly similar to modern jellyfish, with an upper *"umbrella"* dome below which hang large numbers of tentacles.

Analysis of these fossils suggest that these creatures share a number of commonalities with modern cnidarian orders and families. This discovery is significant because jellyfish are sophisticated creatures with complex eyes and nervous systems and advanced mating behavior that includes the ability to recognize potential mates. It appears that these creatures first emerged in something close to their current form as part of the Cambrian explosion.

Fossil remains have also been found of what may have been an apex predator of the Cambrian Period, Anomalocaris. Like the trilobites, this is thought to have been a type of arthropod, though it was gigantic compared to most other forms of life in the Cambrian ocean. Fossil remains suggest that *anomalocaris* (the name means "abnormal shrimp") may have reached one meter in length. It took some time to classify this creature due to the fact that it had both mineralized and unmineralized parts of its body. Its head, for example, seems to have had a hard shell and has left many fossils. However, its more delicate body has left fewer fossils. For this reason, it took paleontologists some time to assemble a complete picture of this creature.

It seems that this was a free-swimming predator that propelled itself by undulating flexible flaps on the side of its body. *Anomalocaris* had an unusual disc-shaped mouth that could close to crush small prey, so it is thought that this creature most likely fed on small trilobites.

All these discoveries tell researchers that the warm, shallow seas of the Cambrian Period teemed with life, some of it very similar to forms still found in Earth's oceans today. Some creatures adopted a benthic lifestyle, foraging for food on the sediments of the seabed while others became nektonic, swimming free in the water column above. Most were invertebrates, but a few, including *Pikaia, Myllokunmingia* and *Haikouichthys,* were the very first kinds of fish and, more importantly, the ancestors of the vertebrates that would eventually come to dominate the planet.

While the Cambrian explosion filled the oceans with life, the lands remained virtually unpopulated. Around 85% of Earth's surface was covered by water (compared to 70% today), and no plant life had yet appeared on land. Even seaweeds are not thought to have existed, or at least no fossil remains have been found from the Cambrian Period.

With that said, land was not entirely devoid of life. Although no plants had yet evolved and no animals were thought to have lived permanently on the land, it is believed that large terrestrial areas were covered in mats of cyanobacteria and other types of microbes. This is not known for certain because these things do not tend to leave good fossil remains, but there is some evidence that the areas where the first creatures were beginning to emerge from the ocean and colonize the land were covered by these mats. No fossils of these creatures have yet been discovered, but there are several trace fossils that appear to show tracks and resting places on these microbial mats. These trace fossils are classified in two distinct groups: *protichnites*, which consist of a furrow flanked by sets of tracks that may have been left by leg-like appendages, and *climactichnites*, tracks left on what were sandy tidal beaches during the Cambrian Period. The best evidence seems to suggest that *euthycarcinoids*, an extinct class of arthropods that may have been amphibious, were most likely responsible for *protichnites*. No creature has yet been identified as being responsible for *climactichnites*, and the best current guess is that these were left by a large, slug-like mollusk. All these trace fossils have been found in areas close to the ocean, which indicates the creatures that left them were perhaps primarily marine inhabitants that occasionally made short trips onto the land.

After the Cambrian Period

The beginning of the Cambrian Period was marked by a mass extinction that killed most species of stromatolites. The warm water pools, where stromatolites had previously thrived, could now be colonized by new types of multicellular animals, including those with carbonaceous shells, leading to carbonate sedimentation.

The Cambrian Period itself was also marked by several mass extinction events. The End-Botomian mass extinction event was actually two separate extinction events that occurred around 513-509 Mya. Scholars still don't know what caused these, though theories suggest changes in sea levels, major changes in the carbon cycle, and possible anoxic events in some ocean environments may have been involved. What is known is that these events caused a major decline in the number of species of trilobites and mollusks.

The next extinction event was the Dresbachian, which took place around 502 Mya. Once again, scientists are not completely certain what caused this event, but, like the preceding End-Botomian event, this extinction killed around 40% of all marine species.

Ultimately, the end of the Cambrian Period was marked by the most significant mass extinction event, the Cambrian–Ordovician extinction event. Approximately 488 Mya, this event drastically reduced the number of trilobite species and completely eliminated some forms of brachiopods and conodonts (eel-like creatures). There have been a number of theories regarding what may have caused this event, including a global reduction in temperature leading to increased glaciation and lowering sea levels or even a massive basalt event, a volcanic eruption that may have covered large areas of the seabed in basalt lava.

Recent studies suggest that the cause may have been anoxia, low levels of oxygen that rendered many parts of the oceans uninhabitable. Analysis of rocks from this period have found very high levels of amounts carbon and sulphur, suggesting that large amounts of both materials were being buried in the seabed. In modern oceans, these high levels are found only in low-oxygen waters like the Black Sea, leading to speculation that large parts of the ocean became anoxic and polluted with high levels of poisonous hydrogen sulphide during the late Cambrian Period. That would certainly account for the steep drop in diversity, though no one has yet been able to determine for sure what might have caused this.

In fact, some scientists have even begun to question whether this extinction event really happened at all, wondering if it might instead be caused by a gap in the fossil record caused by unknown changes in seabed sediments. All that anyone can really be certain about is that the fossil record at the end of the Cambrian Period and the beginning of the following Ordovician had far lower diversity in species and included very few well-preserved fossils.

The Ordovician Period lasted for over 40 million years, from the end of the Cambrian Period (485.4 Mya) to the beginning of the Silurian Period (443.8 Mya). Despite the extinction event with which it may have started, the diversification of complex life forms in the oceans continued in this period - 40 million years after the Cambrian explosion, the *Great Ordovician biodiversification event (also known as the Ordovician Radiation) provided* another explosion of new life and developments of existing species. Invertebrates, mainly mollusks and arthropods, still dominated the oceans, but fish, the first vertebrates, continued to evolve, and the first predatory fish with jaws appeared during this period. Phytoplankton became more common, causing an increase of zooplankton and the development of many suspension feeding creatures. If the Cambrian explosion created many of the modern phyla, the Ordovician Radiation provided another leap in their development and diversity.

However, just like the Cambrian Period, the development of life in the Ordovician Period was almost wholly focused on the oceans. Trace fossils suggest that some creatures may have ventured beyond the water's edge in tidal areas, but there were still no animals adapted to live permanently there. The first known land animal would not appear until the following period, the Silurian.

Pneumodesmus newmani was a millipede-like creature and a fossil dated to 428 Mya has been studied and found to have openings which are thought to be spiracles, part of a gas exchange system that would only work in air. This was a creature that could only live out of the water. However, *Pneumodesmus newmani* was a direct descendent of the arthropods that first appeared during the Cambrian explosion.

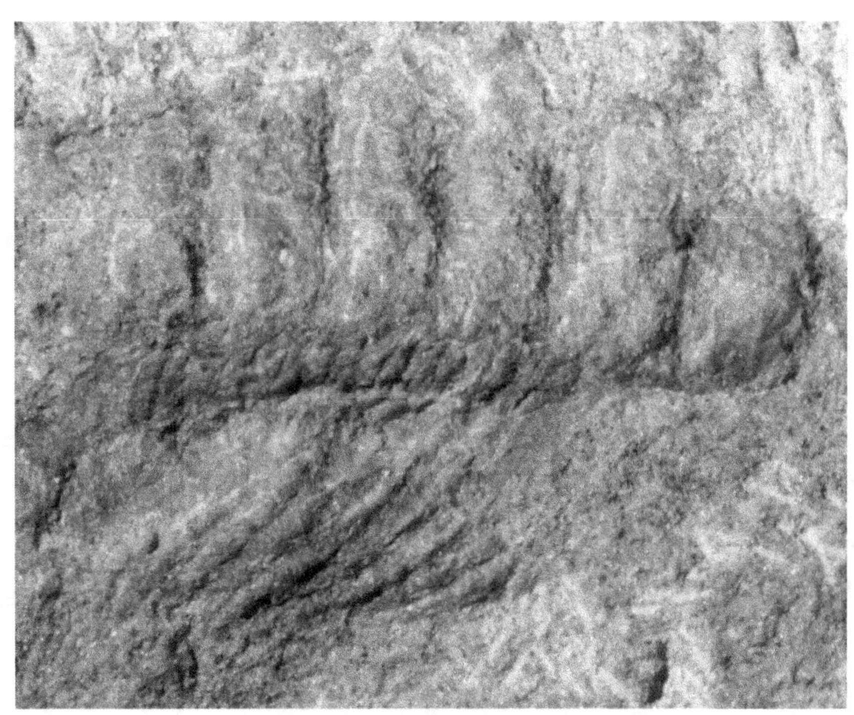

A *Pneumodesmus* fossil

It also seems that the first development of plants took place in the Ordovician, working off of what had been established in the Cambrian. The green algae mats that began to colonize the land in the Cambrian Period seem to have developed in the Ordovician Period into tiny non-vascular forms resembling liverworts. Fossil spores from land plants have been identified in late Ordovician deposits.

People in the 21st century now take the incredible and bewildering diversity of life on Earth for granted, but before the Cambrian Period, no such diversity existed. The Cambrian explosion is well-named, for complex creatures seemed to appear suddenly and everywhere throughout the world, and the fossil record from this period is extensive enough to provide a glimpse into the epoch when it took place. Indeed, almost every animal in the modern world can trace their origins beginnings back to the Cambrian Period.

Online Resources

Other books about ancient history by Charles River Editors

Other books about the Cambrian Period on Amazon

Further Reading

Amthor, J. E.; Grotzinger, John P.; Schröder, Stefan; Bowring, Samuel A.; Ramezani, Jahandar; Martin, Mark W.; Matter, Albert (2003). "Extinction of Cloudina and Namacalathus at the Precambrian-Cambrian boundary in Oman". Geology. 31 (5): 431–434. Bibcode:2003Geo....31..431A. doi:10.1130/0091-7613(2003)031<0431:EOCANA>2.0.CO;2.

Collette, J. H.; Gass, K. C.; Hagadorn, J. W. (2012). "Protichnites eremita unshelled? Experimental model-based neoichnology and new evidence for a euthycarcinoid affinity for this ichnospecies". Journal of Paleontology. 86 (3): 442–454. doi:10.1666/11-056.1. S2CID 129234373.

Collette, J. H.; Hagadorn, J. W. (2010). "Three-dimensionally preserved arthropods from Cambrian

Lagerstatten of Quebec and Wisconsin". Journal of Paleontology. 84 (4): 646–667. doi:10.1666/09-075.1. S2CID 130064618.

Getty, P. R.; Hagadorn, J. W. (2008). "Reinterpretation of Climactichnites Logan 1860 to include subsurface burrows, and erection of Musculopodus for resting traces of the trailmaker". Journal of Paleontology. 82 (6): 1161–1172. doi:10.1666/08-004.1. S2CID 129732925.

Gould, S. J. (1989). Wonderful Life: the Burgess Shale and the Nature of Life. New York: Norton.

Ogg, J. (June 2004). "Overview of Global Boundary Stratotype Sections and Points (GSSPs)". Archived from the original on 23 April 2006. Retrieved 30 April 2006.

Owen, R. (1852). "Description of the impressions and footprints of the Protichnites from the Potsdam sandstone of Canada". Geological Society of London Quarterly Journal. 8 (1–2): 214–225. doi:10.1144/GSL.JGS.1852.008.01-02.26. S2CID 130712914.

Peng, S.; Babcock, L.E.; Cooper, R.A. (2012). "The Cambrian Period" (PDF). The Geologic Time Scale.

Schieber, J.; Bose, P. K.; Eriksson, P. G.; Banerjee, S.; Sarkar, S.; Altermann, W.; Catuneau, O. (2007). Atlas of Microbial Mat Features Preserved within the Clastic Rock

Record. Elsevier. pp. 53–71.

Yochelson, E. L.; Fedonkin, M. A. (1993). "Paleobiology of Climactichnites, and Enigmatic Late Cambrian Fossil". Smithsonian Contributions to Paleobiology. 74 (74): 1–74.

Free Books by Charles River Editors

We have brand new titles available for free most days of the week. To see which of our titles are currently free, click on this link.

Discounted Books by Charles River Editors

We have titles at a discount price of just 99 cents everyday. To see which of our titles are currently 99 cents, click on this link.

www.ingramcontent.com/pod-product-compliance
Lightning Source LLC
Chambersburg PA
CBHW082114170726
47999CB00015BA/3043